EMBROIDERING IDENTITIES

EMBROIDERING IDENTITIES

A CENTURY OF PALESTINIAN CLOTHING

IMAN SACA

in Collaboration with

MAHA SACA

THE ORIENTAL INSTITUTE MUSEUM OF THE UNIVERSITY OF CHICAGO

Library of Congress Control Number 2006936953
ISBN: 1-885923-49-X

The Oriental Institute, Chicago

Oriental Institute Museum Publications No. 25

This volume has been published in conjunction with the exhibition,
Embroidering Identities: A Century of Palestinian Clothing,
held at the Oriental Institute Museum, University of Chicago,
November 11, 2006–March 25, 2007.

About the Author

Iman Saca, PhD, is an Assistant Professor of Anthropology and Director of the Middle Eastern Studies Program at Saint Xavier University, Chicago. She is also a research associate at the Oriental Institute of the University of Chicago.

She specializes in the prehistory of the Levant, and teaches courses on Middle Eastern Culture and Society. Most of her present research focuses on the preservation and protection of endangered cultural and archaeological heritage in conflict areas with a special focus on the West Bank and Gaza Strip. She has presented papers at national and international conferences and co-authored articles on the prehistory of the Levant.

Photo Credits: Object photography by Jean Grant, Monica Hudak, Tom James, and Alison Whyte; pp. 19, 25, 28, and 35 courtesy of the Library of Congress, Prints and Photographs Division; pp. 10, 32, and 39 courtesy of Maha Saca, Palestinian Heritage Center, Bethlehem, Palestine.

Published by The Oriental Institute of the University of Chicago
1155 East 58th Street
Chicago, Illinois 60637 USA
oi.uchicago.edu

Front Cover Illustration: "Royal dress" (*thob al malak*). Bethlehem. OIM A35640A

Cover design by Hanau-Strain Associates, Chicago

Printed by M&G Graphics, Chicago, Illinois.

To my mother, who continually inspires me with her dedication, perseverance, and love

To my father, for his unconditional love, support, and patience

and

To all those who work tirelessly to preserve and promote Palestinian cultural heritage

Thank you

TABLE OF CONTENTS

Jerusalem street scene. Pre-1914. Courtesy of the Oriental Institute Archives, Edgar J. Banks collection

FOREWORD

GEOFF EMBERLING

DIRECTOR, ORIENTAL INSTITUTE MUSEUM

Embroidering Identities: A Century of Palestinian Clothing is a collaboration between the Oriental Institute and the Palestinian Heritage Center in Bethlehem. Together, our collections of hand-embroidered garments are both an evocation of a particular time and place and a meditation on how traditions are constructed, maintained, and remembered. The historical setting is the end of the Ottoman empire and the time of the British Protectorate, and more specifically, the villages of Palestine from about 1880 to 1948. Working-class women in those villages, having perhaps more free time than spare money, developed locally distinctive ways of making and embellishing dresses and head coverings. Passed from mother to daughter, these embroidered patterns developed into local styles that were maintained for the better part of a century. After a lapse of decades, these traditions are being re-invigorated as a symbol of the Palestinian past.

A display of early modern Middle Eastern clothing represents a departure for the Oriental Institute. With the inauguration of the new Marshall and Doris Holleb Family Gallery for Special Exhibits, we are able to broaden the range of material on display in the museum. In this case, concerns of archaeologists and historians — like style, craft production, and tradition — are illustrated by more recent objects from the Middle East. Our collection of Palestinian garments was donated to the museum in 1980 by Mrs. Clara Klingeman, having been purchased in the late nineteenth and early twentieth century by John Whiting for the American Colony Store.

This exhibit would not have been possible without the curatorial vision of Iman Saca and the many contributions of Maha Saca of the Palestinian Heritage Center. It was the hard work and dedication of Oriental Institute staff that brought the exhibit to life in complex circumstances and under the time pressures that resulted: Emily Teeter (Coordinator), Tom James (Curatorial Assistant), Erik Lindahl (Head Preparator), Brian Zimerle (Preparator), Helen McDonald (Registrar), Laura D'Alessandro (Head of Conservation), Alison Whyte (Assistant Conservator), Monica Hudak (Contract Conservator), John Larson (Archivist), Jean Grant (Photographer), Denise Browning (Museum Shop), and the Oriental Institute Publications Office including Tom Urban (Senior Editor), Leslie Schramer (Assistant Editor), Katie L. Johnson (Editorial Assistant), and Lindsay DeCarlo (Editorial Assistant). Thanks also to Wadad Kadi whose special expertise was so useful. The exhibit graphics were designed with flair by Dianne Hanau-Strain. Special thanks to Candace Walters in the Office of Risk Management of the University of Chicago, who worked patiently on the unusual complexities of insuring this exhibit from start to finish.

Myriam Nasrallah Khamis (b. 1894) wearing a traditional Bethlehem dress (*'anbar thob*) and Bethlehem headdress (*shatweh*) covered with a white veil. Photo courtesy of Maha Saca

PREFACE

Soul searching for one's identity and heritage comes in different ways. In 1990, my mother recognized that the younger generation of Palestinians was struggling to find a way to relate to their heritage. Their cultural heritage is rich, but it was only known to the elders and a few individuals with personal interest in the subject. There were very few efforts to record traditions, songs, stories, or the art of embroidery. They were all disappearing.

Realizing the difficulties of preserving the rich cultural heritage of Palestine and promoting it, especially to the younger generation of Palestine, my mother embarked on a mission of collecting and documenting folkloric stories and traditional items such as embroidered clothing, jewelry, and traditional household items.

She conducted fieldwork in many villages and refugee camps in the West Bank, gathering much information mainly from women. She asked them about stories from the villages they left, the names of traditional household items they used, and basic farming tools. Most importantly, she asked them about their wedding day, about their *thobs*, the wedding procession, the women's bridewealth, the songs they sang, and the wedding ceremony. These women spoke candidly about their experiences. They brought out their dusty wedding *thobs* and they told the story of its production and the meaning behind the colors and the motifs. These marvelously embroidered pieces told an amazing life story, a story that would have disappeared if there was no interest or effort to record it.

During the time of her research, I was studying in the United States. Every time I called she would relate the stories she collected. I was excited to hear stories, but most importantly, I was excited to hear the excitement in my mother's voice. I became very passionate about the work that she does and, in 1991, my mother and I started the Palestinian Heritage Center (PHC) in Bethlehem close to the Nativity Church.

Today, the Palestinian Heritage Center is considered one of the most important centers for the preservation and promotion of Palestinian heritage. The center includes a library that is used by visitors and students. My mother and I constantly give presentations to schools and college students who visit the center. The center also has a complete collection of traditionally embroidered Palestinian dresses. One of the important activities of the center is to teach embroidery to young Palestinian women who are becoming very interested in learning the skills of their grandmothers. The center also provides job opportunities for sixty-five women from various villages around Bethlehem. Many of these women embroider on modern clothing which enables the young generation to wear contemporary clothing, yet preserves the beauty of the traditional Palestinian embroidery.

The beauty, depth, and richness of the traditional embroidery will stay alive due to the efforts of many committed individuals ... Like my mother, Maha Saca.

Iman Saca
Exhibit Curator
Chicago, 2006

PALESTINE
Before 1948
MEDITERRANEAN SEA
LAKE TIBERIAS
JORDAN RIVER
Tulkarm
Jaffa
Beit Dajan
Ramallah
Jericho
Mejdel
Jerusalem
Bethlehem
Beit Jebrin
Gaza
Hebron
DEAD SEA
Bir Sabe'

INTRODUCTION

Clothing expresses aspects of identity in all societies. In Palestine of the nineteenth and early twentieth centuries, handmade and richly embroidered women's garments expressed regional identity at the same time as they marked age and status. These dresses and headdresses were made mainly by villagers and Bedouin (*Bedu*) and not made or worn by the urban elite. In contrast to the marked regional diversity in women's clothing, men's clothing was generally similar throughout the different regions of Palestine and in Syria, Lebanon, and Jordan. The traditional male ensemble was made up of a coat (*qumbaz*), another type of overcoat (*'abaya*), a headdress (*keffiyeh*), and a belt (*hizam*).

Before the middle of the twentieth century, women in each local region created garments with distinctive types of embroidery and decoration that immediately established the wearer's origin. To those who knew the regional variations in style, patterns, and colors of embroidery, a quick look at a dress was enough to determine the wearer's region and even village. Marital status was also expressed through specific styles of garments that distinguished unmarried girls, married women, widows, and women who wished to remarry. *Thobs* (the generic term for a dress-like garment) and headdresses were also excellent communicators of wealth and social standing. Today, such traditional distinctions of dress have almost vanished due to both the political circumstances that have changed the social and economic life in Palestine and the new patterns of socialization and increased communication. Distinctive regional garments have been largely replaced by modern Western clothing that reveal nothing about the wearer's origins.

Palestinian girls began learning embroidery and dressmaking skills from their grandmothers at about the age of seven. From this early age a girl began creating items for her wedding trousseau. By the time she married, the bride's trousseau included many lavishly embroidered items. People strongly believed that the personality of the future bride was revealed in the workmanship, color, and design of her dresses. These garments remain among the best-known traditional handicrafts of Palestine. Men's clothing, on the other hand, had little social significance and no elaborate embroidery.

BACKGROUND

The population of Palestine inhabited different geographical zones: the coastal regions, the central highlands, the Jordan River Valley to the east, and the semi-desert areas of southern Palestine. These areas were ruled by the Ottoman empire until the end of World War I and by the British Mandate from 1918 to 1948. Unfortunately, the northern region of Palestine, which includes Galilee, Nablus, Tulkarm, and Jenin, is not incorporated in this exhibit due to the lack of representative dresses in our collection.

The presence of foreign cultures naturally affected the styles of dress in Palestine. In general, townspeople were strongly influenced by the introduction of new clothing styles and materials, whereas people who lived in villages borrowed sparingly from foreign styles and gave a local touch to what they adopted from European travelers and pilgrims. The fashions of the Ottoman Turks had a particularly strong influence on women's dress in northern towns, especially among the ruling class.

Although foreign elements were borrowed during the period of Ottoman rule and British Mandate, regional distinctions in village and Bedouin dress design and embroidery styles remained strong; one could still clearly distinguish which village a woman came from by her dress or headdress. The embroidery on her clothing also helped the viewer determine the woman's age, marital status, and wealth.

The establishment of the state of Israel in 1948 saw a radical social, economic, and political transformation of the region. Many Palestinian villages were destroyed, abandoned, or occupied, and people sought refuge in neighboring countries and in the West Bank and Gaza. Life for many Palestinians was disrupted and many aspects of their society and culture were affected. Traditional modes of dress and the art of embroidery were impacted, for many women could no longer afford the time or the money to embroider the complex garments. Despite

"Royal dress" (*thob al malak*) with short velvet jacket (*taqsireh*). Bethlehem. OIM A35640A, A35652

the political and economic difficulties, however, women in the refugee camps, mainly in Jordan and Lebanon, continued to embroider in the style of their original villages in an attempt to maintain their displaced identity. By continuing the tradition of embroidery and wearing traditional dresses, women felt that in their own way they were keeping part of their heritage and village alive. Even today, there are individuals and many women's cooperatives and centers that work to safeguard this tradition, develop it, and teach it to the younger generation.

DRESS TERMINOLOGY

Traditional Palestinian dresses are composed of different elements that were assembled after they were embroidered or decorated. The different parts of the dress were joined together by two main stitching techniques: the *manajel* and *sanabel* stitches.

• Chest panel (*qabbah*): This is the most important and elaborately embroidered part of the dress. In most regions, the chest panel is embroidered separately then attached to the dress. The color and motif of the chest panel differ from village to village and are therefore clear indicators of where the wearer comes from.

• Sleeves are of two main types and are also regionally distinct, clearly identifying the villages or regions from which the dresses come.

Irdan are long pointed triangular sleeves. The pointed ends of the *irdan* are usually left unembroidered because women tied this part of the sleeve behind their back when they worked in the field or around the house. In some areas like Jericho, women actually used the *irdan* to help them carry heavy objects and, in some instances, they were used to cover the head.

Kum are narrow sleeves that can be either long or short. The embroidery on *kum* is executed mainly on the outer part of the sleeve in vertical bands of varying thickness that start at the shoulder and end near the cuff stitch. The area around the cuff may also be heavily embroidered and strips of satin might be added to the sleeve.

• Shoulder piece or yoke (*radah*): Depending upon the region, the *radah* could be made of striped atlas silk, satin, or velvet embroidered with elaborate geometric motifs. The *radah* stretches from the middle of the shoulders all the way to the upper part of the chest, where it was usually attached to the top of the chest panel. The size of the *radah* varies by region and serves several different functions. Mainly, this piece adds to the beauty of the dress by allowing women to use a variety of materials and designs that reflect their personal taste or village identity. It also protects the dress from hair dyes such as henna, and it prolongs the life of the dress by providing extra protection from wear and tear around the area at the neck and shoulder.

• Skirt side panels (*benayiq*): These triangular pieces of fabric make up the sides of the dress. Some side panels are made from the same material as the dress, while others are of a different material to draw attention to this area of the garment. The *benayiq* have a practical purpose; they widen the dress so a woman can walk comfortably, since most dresses do not have a side or back vent. The *benayiq* of the Bethlehem bridal dress are famous for their beautiful motifs embroidered with gold and silver threads and its three vertical silk strips which were often made of a combination of colors like red and green. Women from different areas often bought these side panels separately and sewed them onto their dresses.

• Front skirt of the dress (*hijjer*): This part of the dress shows much regional variation. Some dresses, for example, the Bethlehem everyday dress, have plain fronts with no embroidery. In other regions, the *hijjer* of the dress is heavily embellished with embroidery and decorative patches of silk and velvet. Wedding dresses from the Hebron area are of this latter type. Most dresses, however, have two elongated bands

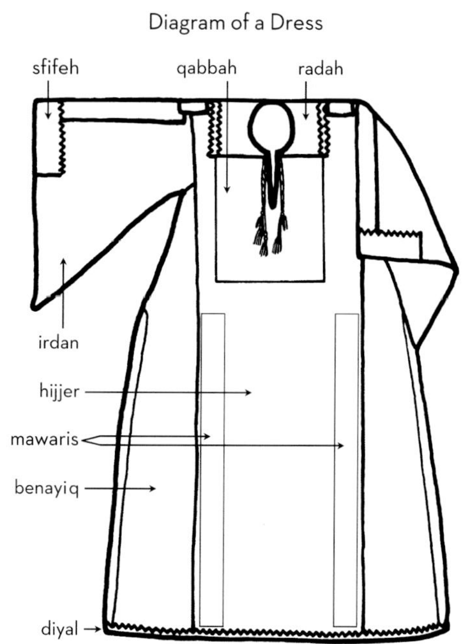

of embroidery (*mawaris*) that start at the waist and end at the hem of the dress on the left and right sides of the *hijjer*.

• Back lower panel (*shinyar*): This part of the dress is heavily embroidered in many regions. The back panel, like the front, may be framed by two *mawaris* strips. The *shinyar* is an important part of the dress, for it is a showcase for the needlepoint skills of the woman who created it. It was the part that was most admired and critiqued.

• Hem (*diyal*): The embroidery on the hem could be narrow or broad depending on the region. The *diyal* helped protect the bottom edge of the dress from the wear and tear of everyday activities.

FABRICS AND DYES

The fabrics used to produce dresses are primarily linen, cotton, wool, and silk. These fabrics were mostly handwoven at large weaving centers like Mejdel, Bethlehem, Nablus, and Ramallah. The raw material for these fabrics was either produced locally in Palestine or imported, especially from Egypt and Syria. Many of the high-end fabrics were also imported from Europe, India, and other Mediterranean countries. The famous silk/cotton satin called atlas silk was imported from Damascus, and the *rumi* linen used for Ramallah dresses was mostly obtained from Greece and sometimes Turkey.

Some fabrics have specific names. For example, linen with a higher percentage of silk than other fabrics is called *malak* ("royal") and was used for the Bethlehem bridal dress. It was woven in Bethlehem and in the main weaving center in Mejdel. *Biltajeh* is an indigo-dyed cotton fabric with colored silk borders. It was handwoven in Mejdel and was used mainly for the dresses of the coastal region. *Jinneh u nar* ("heaven and hell") is a silk fabric made of red and green vertical stripes; the color red symbolizes hell and green symbolizes heaven. This fabric was mostly used for bridal dresses in Bethlehem and Jerusalem and was produced mainly in Mejdel. *Heremzi*, a narrow silk fabric woven in Syria, was used for the side panels and the sleeves of many dresses. *Ghabani* is a white silk fabric embroidered with yellow silk thread that was produced in Syria and used in the Jerusalem *ghabani* dress, named after the material.

Men tended to favor striped atlas silk, or "roza silk," for the overcoat (*qumbaz*), and for the sleeveless overcoat (*'abaya*) they preferred soft wool or silk.

The two traditional colors of fabrics used for Palestinian dresses were un-dyed linen and indigo blue, the favorite color of most villages and Bedouin. White linen or cotton fabrics were favored and used in some villages like Ramallah and Beit Dajan.

Linen shawl (*khirkah*) embroidered with S-shaped "leech" (left and right borders), tall palms (center), eight-pointed star (left center), and feathers (left and right edges and above palms) patterns in silk thread. Ramallah. OIM A35635B

Dyes were produced from local plants and insects. Producing these colors was the specialty of a few families in Palestine, and the production processes were kept as very well-guarded family secrets. The use of synthetic dyes began in the late nineteenth or early twentieth century, but locally produced natural colors continued to be used. Natural dyes were more attractive to most women because their color stayed vibrant and did not fade after many years of use.

Sources of Natural Dyes:

- Red was produced by mixing pomegranate skins with other plants (for example, madder and kermes) or insects (cochineal).

- Indigo, in many different shades of blue, was made from a plant (*Indigofera tinctoria*) cultivated in the Jordan River Valley.

- Yellow-green was produced by the sumac plant.

- Purple was produced from crushed murex shells.

- Black was produced using walnut skins.

- Yellow was produced from grape leaves.

Many villages and even entire regions were associated with the use of certain color fabrics and threads. For example, women from the region around Ramallah and, to some extent, Jaffa, favored thread of various shades of red, especially wine-red, as the main color for embroidering their dresses. Women from Hebron preferred browns and red-browns.

EMBROIDERY PATTERNS AND MOTIFS

Patterns and motifs are the hallmark of Palestinian embroidery. Motifs were of utmost importance, for the way a woman mixed colors and designs made a statement about her identity and skill. A standard repertoire of designs and motifs are repeated in realistic or stylized patterns in a variety of colors. Geometric shapes such as circles, rectangles, or squares, and various floral and foliage patterns were favored in the late nineteenth–early twentieth century. These designs are few in number, but women created many combinations of them.

Specific motifs were given names and meanings. For example, the eight-pointed star of Bethlehem is named for a meeting between the moon and Astarte, the Canaanite goddess of fertility. Motif names often reflect ordinary items from a rural woman's daily life, such as food (apples and chick-peas), animals (cow's eye and scorpions), and implements (mill wheel and ladder). An S-shaped pattern was also known as the horse's head or leech. Other motifs symbolized basic elements in nature like the sun, moon, stars, trees, mountains, and water.

Embroidered motifs could express a woman's hopes and expectations, and even her attitude toward certain issues such as marriage, children, and in-laws. If a woman from Hebron, for example, wanted to have children, she might embroider "doll" figures on the back panel of her dress to express this desire. In social gatherings with other women from adjacent villages or towns, women had no need to ask where the other women came from, for their placcs of origin wcre apparent in the dresses they wore. For example, women from Beit Dajan often embroidered an orange blossom on their dresses, for this village was known for its orange orchards. Cypress trees embroidered between the orange blossom motifs symbolized the boundary between the orange orchards. A careful observer, however, would be able to recognize if a woman wearing an orange blossom dress advertising her residence in Beit Dajan was originally from a different village by taking note of motifs and stitches that were not typical of Beit Dajan.

"Cypress tree" (top) and Pasha's tent (bottom) motifs. Jaffa-Beit Dajan. OIM 35643A

Border patterns are also an essential part of Palestinian dresses. These can be a major part of the design, for example, the embroidered branches (*mawaris*) on the front and back skirt of the dress, or they can frame other motifs on the garment. The most famous of these border motifs are the feather motif, toothed lines, comb pattern, ladder, roses, soap, and triangles.

Local designs and motifs were greatly influenced by the constant passage of pilgrims and merchants through Palestine and by the many foreign peoples that visited and stayed in the country, especially in the early twentieth century. Palestinian women would copy or emulate patterns they saw on rugs, tiles, carpets, and in magazines brought from other lands.

EMBROIDERY COLORS

One of the most attractive aspects of Palestinian embroidery is the variety of colors and the arrangement of color schemes. Red, the color of happiness and life-blood, was used in almost all Palestinian embroidery in the late nineteenth–early twentieth century. It was used on white, dark blue, and black linen or cotton fabrics. Village and Bedouin women were attuned to the various shades of red and they combined them in spectacular ways. Many different colors, such as yellow, green, pink, orange, and white, were used to accentuate the red tones. The elaborate silks that produced these brilliant colors were mostly imported from Greater Syria as well as from Europe. Starting from the late 1930s, cotton threads — imported from France — replaced silk threads.

Color schemes were created in a manner that expressed a woman's feeling and stage in life. For example, in the Hebron region, older women wore dresses embroidered with purple threads whereas younger women and girls embroidered their dresses with red and green. In some Bedouin tribes, unmarried girls wore dresses embroidered on the back panel and the front panel with blue, while married women used the color red to signal their status as

Chest panel (*qabbah*) embroidered with vase pattern (central area) surrounded by the chevron-shaped *sabal* border. Bethlehem. OIM A35640A

Gold cord embroidery (couching) in floral pattern on the front of a short jacket (*taqsireh*). Bethlehem. OIM A35652

wives. Older women or widows who were interested in remarrying combined the color blue with red flowers and sometimes intertwined figures of children with blue embroidery on the back panel of their dresses.

EMBROIDERY STITCHES

Of the numerous embroidery stitches used in Palestine, two stand out more than all the others: the cross-stitch and the couching stitch.

Cross-stitching was most commonly used in the central, southern, and coastal regions of the country. Most of the nineteenth and early twentieth century traditional dresses were made with a coarse weave that facilitated counting the threads of the warp and the weft, enabling women to cross-stitch on the fabric directly without a pattern. With the introduction of modern fabrics, women used embroidery canvas or a pattern since most new fabrics are too tightly woven to allow for the easy counting of threads.

Couching is a technique in which cords were laid down on the fabric and secured with tiny stitches. Women in Bethlehem were famous for couching gold and silver cords, and since most of the embroidery on wedding *thobs* was done on fine-textured silk fabrics, it was easier to embroider using a couching stitch. Women throughout Palestine traveled to Bethlehem to purchase chest panels embroidered with the couching stitch. Dresses and panels employing the couching stitch were unique, festive, and beautiful. Villagers felt that they were buying a premium product from the town of Bethlehem, a product that was created by women who specialized in this kind of work.

Other stitches used in Palestine are the hem stitch, satin stitch used as a filler to give solid appearance, *manajel* and *sanabel* stitches used to hold selvedges together, chain stitch, and stem stitch.

REGIONAL DRESSES OF PALESTINE

The style of traditional clothing in Palestine was a reflection of geography and the individual's economic status, descent, aspirations, and lifestyle. Combined, these created the inspirational base for these village dwellers.

In larger towns, clothing styles were highly influenced by fashions imported from the Ottoman empire and Europe. Townsfolk had continuous interactions with British and Ottoman officials and well-to-do individuals who resided in the larger cities of Palestine. The clothing that townswomen wore reflected this influence. Imported dresses and pieces of imported fabric lightly embroidered with metal threads were worn in large Palestinian towns like Jerusalem, Nablus, and Jaffa; the fabric and metallic embroidery were considered prestigious and fashionable.

In contrast, the Bedouin women wore dresses that are characterized by heavy embroidery on the back and front panels and the lack of embroidery on the long pointed sleeves. The dress panels are embroidered with colors and motifs that reflect the women's social and marital status as well as their beliefs and aspirations. For example, widows who wanted to remarry embroidered colorful motifs on the back panel.

Woman wearing the Bethlehem "royal dress" (*thob al malak*) with a short jacket (*taqsireh*) and a coin-ornamented headdress (*shatweh*) from which hangs the *iznaq*. The headdress is partially covered by a white veil. Bethlehem. Library of Congress, Prints and Photographs Division, LC-DIG-matpc-11764

Villagers made up the majority of the population in Palestine. Dress styles and embroidery designs were similar in villages of the same region. Villagers did not travel for long distances nor did they see the need to do so.

Village women were the main producers of many handcrafts including basketry, pottery, and, of course, embroidery. Through their dresses, these women expressed their sorrows and joys, their beliefs, and their relationship with nature. Village and Bedouin women competed among themselves to embroider the best and most beautiful dresses. But if we look closely, most of the garments contain mistakes, either in color or motif. Women made these "errors" intentionally in order to ward off envy by not creating a "perfectly" embroidered *thob*.

Since marriage was a very important event in Palestinian society, much activity revolved around preparation for the wedding ceremony and the wedding day. Many of the *thobs* are wedding dresses or dresses that were part of a girl's wedding trousseau.

THE COASTAL REGION

Jaffa Area-Beit Dajan

In the Jaffa area, dating back to the early–mid-nineteenth century, the *jallayeh* was the most important dress of a bride's trousseau. It is usually made from indigo linen and is heavily embroidered on the front skirt, lower back panel, and short narrow sleeves. The front of the skirt is open, and the opening is heavily decorated with patches of red, yellow, and green taffeta silk; the shoulder yoke is covered in red and yellow striped atlas silk. The elaborate cross-stitch embroidery is composed of amulets (*hijab*), arches, and cypress trees (*saru*). The decoration of the chest panel echoes some of the motifs on the dress, with special attention given to cypress trees and amulets. The sleeves are short and heavily decorated, and a strip of red and yellow atlas satin frames the ends of the sleeves. Tassels of pearl and silk hang from the opening at the neck.

Dress of the Jaffa-Beit Dajan area with taffeta patchwork along the opening of the skirt. The front is embroidered with the triangular amulet, eight-pointed star, feathers, and inverted cypress tree patterns. The back panel (*shinyar*) is very heavily decorated with the Pasha's tent and cypress trees motifs. OIM A35643A

The lower back panel is the most elaborate, noticed, and critiqued. On this garment it is heavily embroidered with mainly dark orange silk thread accented with purple, yellow, and green. The embroidery panel is so large that it covers the back of the garment. As with the rest of the *jallayeh*, amulets, arches, and inverted cypress trees are the main motifs. The density of embroidery and the care with which it was created make this *jallayeh* one of the best known of all the dresses from the area.

Headdresses (*smadeh*) of the Jaffa region are heavily embroidered with cross-stitch and have a row of coins attached at the front. Various kinds of veils were used in this area. Black or red *shambar* veils embroidered with silk thread are early examples. In the early twentieth century, many women wore colored veils with flower prints or simple rayon veils, similar to the ones worn in Ramallah.

GAZA

The dresses that originate from the area around the port city of Gaza are made of cotton and silk fabric woven in the town of Mejdel on a traditional loom (*nol*). The fabric was woven in strips (*maqta'*) in a specific size: 33 centimeters in width and 7 meters in length. These strips were then connected to one another using the *manajel* stitch. The fabric of the Gaza *thob* was given various names based on its fabric, design, and color. *Abu Mitain* is indigo blue cotton with wide magenta strips and smaller red and green strips on each side of the dress. The *Abu Mitain* dress — also the name of the fabric — lacks embroidery although sometimes a small embroidered amulet is sewn at the bottom of the neck opening.

Abu Mitain thob with triangular chest panel. The side panels are embroidered with amulet motifs. Gaza. PHC 165

The *Abu Rukbeh* dress is made of indigo blue cotton with heavy silk embroidery on the chest panel. In contrast to chest panels of other regions, the Gaza *Abu Rukbeh* dress has a triangular-shaped chest panel embroidered in vibrant colors, mostly violet-red, purple, and sometimes blue and green, depending on the dressmaker's choice. The narrow sleeves usually have little or no embroidery. Both sides of the dress are heavily embroidered from the waist down with various motifs especially the amulet *hijab* motif.

The *zuniyet* dress is also made of indigo blue cotton striped with magenta and light green. The square chest piece echoes those from other regions in Palestine, but it is quite different from the usual triangular chest panel of Gaza and the southern coastal region in general. The dress is heavily embroidered with amulets and triangles on the front and back panels and on its narrow linen sleeves.

The most common headdress worn in the Gaza region was the beaded cap (*takiyat shabak*) made of colorful, especially blue, beads. Some of the *takiyas* have coins sewn on them in an attempt to imitate the more elaborate *smadeh* of the other coastal villages.

The head veil (*gudfeh*) is made of cotton or silk and is embroidered with colorful silk threads. The ends of the head veil are lined on all sides with small colorful beads.

THE CENTRAL REGION

JERUSALEM

Even though Jerusalem was a major city, it never developed its own characteristic style of embroidery. A likely explanation for this is that it was a vibrant, international city and was constantly exposed to foreign influences as people from all over the world brought their own regional and national costumes.

Dresses of the Jerusalem area were mainly made of fabrics imported from Damascus. These fabrics were made of silk and were very colorful, which made them very distinct from the indigo blue or black dresses that dominated other areas. Some of the fabrics imported into Palestine from Syria were the *asawri*, *heremzi*, atlas, and *ghabani*. *Asawri* is silk fabric vertically striped in either red and yellow or in yellow and black. *Ghabani* silk is yellow with yellow silk threads delicately overlain in beautiful patterns. A woman prepared many embroidered dresses for her wedding trousseau, including dresses made of *asawri* and atlas silk. The *ghabani* dress was the most important, as it was the actual wedding dress.

Fashion in Jerusalem was influenced by dress styles of the surrounding towns and villages, particularly Bethlehem. Jerusalem dresses were often a mixture of various regional styles. A woman might have borrowed the Bethlehem chest and side panels and yoke and sewn them onto Damascus silk fabric. Since the women of Bethlehem and Beit Jala were very well known for their needle-point skills, they were often commissioned by women from Jerusalem to embroider dresses with the famous Bethlehem chest and side panels and even the yoke. The shoulder yoke was made of velvet and silk, and, as in Bethlehem, featured the bird of life (*ta'er al finiqe*) or floral motifs among its embroidered designs. These were applied with gold thread using the couching technique.

In the early twentieth century, velvet imported from Europe became popular, especially in red, blue, and green. The use of velvet emphasized the "higher" status of Jerusalemite women and distinguished their garments from all other Palestinian dresses.

In contrast to other elaborate headdresses in parts of Palestine, the Jerusalem headpiece (*qub'a*) is much simpler and lacks elaborate adornments. The *qub'a* is similar to that of Bethlehem's round cotton head cap. The *qub'a* is embroidered in cross-stitch and more elaborate headdresses are embroidered in gold couching stitch and have coins attached to their edges. Some women from Jerusalem also wore the Bethlehem *shatweh*.

In some areas of Jerusalem in the late nineteenth century, women wore an embroidered head veil (*ghudfeh*). Later, most head veils were made of simple white cloth made of silk with no ornamentation or embroidery.

Belts (*hizam*) varied and some were made of atlas silk and are similar to the belts worn in Ramallah.

Jerusalem *thob* made of *ghabani* silk. The chest panel (*qabbah*) and side panels (*benayiq*) are from Bethlehem. The tops of the sleeves and side panels are decorated with the Bethlehem watches (*sa'at*) pattern. PHC 119

RAMALLAH

Throughout Palestine, Ramallah was famous for its handwoven, white *rumi*-linen embroidered *thob* with cross-stitching in wine-red silk. Black linen *thobs* were also used in Ramallah, but they were mostly worn on special occasions and in the winter months. White *rumi* are considered summer *thobs*, but women could wear them year round.

The chest panel of the Ramallah *thob* is decorated with a distinctive arch *qos* motif. The distinct Ramallah tall palm motif is heavily embroidered on the back lower panel of the dress. The vertical bands (*mawaris*) on the back and front of the dress including stars (*nujum*), S-shapes (called horse's heads or leeches), cypress trees (*saru*), and feathers (*reesh*).

The sleeves of the Ramallah *thob* were either narrow with embroidery on most parts of the sleeve or were triangular and pointed *irdan* with embroidery on only the upper part of the sleeve and the cuff. The narrow sleeve

White *rumi thob* with atlas silk belt (*zunnar*). Ramallah. OIM A35635A

Black *rumi thob* with skirt and chest panel heavily embroidered in the eight-pointed star motif. Ramallah. OIM A35633A

was more popular in Ramallah during the late nineteenth–early twentieth century. Later, the *irdan* became more fashionable, especially for dresses worn on special occasions.

The distinctive Ramallah headdress, known as the *smadeh* or *saffeh*, is shaped like a horseshoe and is heavily embroidered in cross-stitch with floral and small geometric motifs. The sides of the headdress are padded and rolled so the heavy row of closely arranged coins can be easily and comfortably attached to the headdress. A silver chain (*iznaq zarad*) with a silver or gold Maria Theresa coin, the most valued coin in the region, suspended from its center hangs from the side of the headdress. An extension (*joseh*) could be added to the lower part of the chain to make it appear longer and fuller. For wealthier women, another horseshoe-shaped band (*toq*) of Maria Theresa coins was placed on top of the original *saffeh*.

Ramallah woman wearing the linen *rumi thob*, her coin-rimmed headdress (*saffeh*) partially covered by an embroidered veil (*khirkah*). Circa 1898–1946. Library of Congress, Prints and Photographs Division, LC-DIG-matpc-11836

The head veil (*khirkah*) is made of the same material as the dress and is heavily embroidered with beautiful red-colored thread. The *khirkah* is usually made of two separate pieces of linen joined in the middle with tassels attached to each end. Medium-sized horizontal bands were embroidered on the long sides of the veil while very wide heavily embroidered bands with tall palm motif were created on the shorter sides of the veil. Earlier head veils were embroidered with multi-colored silk threads while headdresses dating to the early twentieth century were embroidered with various shades of red and black silk threads, but the same basic structure of the older head veils was retained. The tall palm trunk was the predominant motif on both these veil types.

Around the mid-twentieth century, some women exchanged the traditional *khirkah* for a large silk shawl (*shal*) that was first imported from Europe but later made locally in Ramallah. These veils were mostly magenta in color and machine embroidered with large floral motifs. They also had very long tassels made of the same color and material as the rest of the shawl.

The belt (*zunnar*) of the Ramallah dress was made of burgundy and yellow atlas fabric. Some belts were decorated with tassels at both ends.

Bethlehem

The Bethlehem royal dress or *thob al malak* is one of the most distinctive dresses of Palestine. During the nineteenth and early twentieth centuries it became fashionable for women from all over Palestine to purchase the *thob al malak*. For women who could not afford the entire dress, the chest panel sufficed. The Bethlehem dress was unique and distinct, especially since it was embellished with gold and silver and silk threads. This beautiful

Thob al malak (bridal gown). The watch pattern (*sa'at*) is couched on the top of the sleeves and on the side panels (*benayiq*). The chest panel (*qabbah*) has a vase pattern surrounded by the *sabal* herringbone stitch. Bethlehem. OIM A35640A

dress was worn by the bride on her wedding day and on subsequent special days.

There are two kinds of bridal dresses in Bethlehem: the *thob al malak* and the *thob al malak al ikdari*. The dresses are similar except for the color of the fabric. The *thob al malak* is made of linen and silk with vertical stripes in varying shades of red. The linen fabric contains a high percentage of silk and required extensive labor and was therefore one of the most expensive fabrics in the region. The *thob al malak al ikdari*, which translates as "green royal dress," is made of red, green, and indigo striped fabric. Like the *thob al malak*, the *thob al malak al ikdari* is made of fabric with a high percentage of silk and required extensive labor.

Shatweh (married woman's headdress) adorned with bridewealth coins. Detail on right. Bethlehem. OIM A35640E

The *thob al malak* is embroidered with the distinctive Bethlehem couching in silk, gold, and silver threads in circular and zigzag patterns. Couching in silk, which began in the nineteenth century, is known as *tahriri*, and the use of gold or silver cord, which started in the early twentieth century, was referred to as *qasab*. This decoration gave the Bethlehem embroidery style a unique appearance and made it very popular in other villages and towns throughout Palestine. Women from the Bethlehem region were often commissioned by residents from other towns and villages in Palestine to create the wedding day dress, or at least the chest panel. Along with Beit Jala and Beit Sahur, Bethlehem was a largely Christian community, so women were familiar with the elaborately decorated robes of the patriarchs and church furnishings, as well as with ornate Ottoman clothing. These influences probably contributed to the introduction of the couching technique in Palestine, especially the use of gold and silver threads.

Velvet jacket (*taqsireh*) with gold couching decoration. Bethlehem. OIM A35652

Sleeveless striped wool cloak (*bisht*). Bethlehem. OIM A35646B

The chest and side panels were the most decorative parts of the Bethlehem dress. The couching is done on red or green taffeta, and the borders are done in colorful green, yellow, or red, mostly in triangular appliqué. In the early twentieth century, Bethlehem women became so sophisticated in creating *tahriri* and *qasab* couching that the taffeta background became obscured as the embroidery became more elaborate. The triangular taffeta side panels of the *thob al malak* are heavily decorated. They are usually divided into three sections with vertical *tahriri* embroidery. The bottom of the panel is decorated with couching of gold and silver of the most representative design of the Bethlehem region: the watches (*sa'at*) design.

Another very distinctive characteristic of the Bethlehem bridal dress is the short jacket (*taqsireh*) worn over the *thob*. Originally, it was made of broadcloth embroidered with *tahriri* silk cord couching. The *taqsireh* had short sleeves below which were exposed the long triangular sleeves of the *thob* with their heavily couched cuffs. Over time, the *taqsireh* became more intricate with elaborate gold and silver couching done on velvet instead of broadcloth.

The Bethlehem hat (*shatweh*) was worn by married women. It was made from padded and lined broadcloth. The *shatweh* was embroidered across the top using cross-stitch and heavily embroidered down the sides with gold and silk couching. It was adorned on the front panel with coins that were part of a woman's bridewealth. Pieces of coral may have been attached to the front of the headdress. Unmarried women wore a heavily embroidered circular cap (*qub'ah*), some of which were also adorned with coins. This cap facilitated the attachment of the white veil that covered the woman's head and hair.

Bethlehem veils in the nineteenth century were elaborately embroidered on cotton or linen fabric. The famous Bethlehem star motif (*najmah kin'aniyah*), the feather (*reesh*) motif, and the *sanabel* stitch were embroidered on all four sides of the head veil. Around the mid-twentieth century, simple white veils were worn by most women.

Everyday Bethlehem *thobs* (*'ambar*) were not as lavishly embroidered and were generally made of black fabric. The main decoration was cross-stitched embroidery on the chest panel, and the sides were embroidered with the *manajel* stitch. Over the *thob*, women (as well as men but in different colors) wore a sleeveless coat (*bisht*) made from locally spun and dyed wool. The narrow, vertical woven strips of red and black wool end at a heavily couched thick border at the hem and the shoulder. The *bisht* was mostly worn during the winter and cold evenings.

Woman wearing a voluminous bloused Jericho dress. Circa 1940–1946. Library of Congress, Prints and Photographs Division, LC-DIG-matpc-12962

The belt (*shamleh*) worn by Bethlehem women was made of locally woven wool fabric in pink and blue. The color pink, like the color red, was mostly worn by married women, while blue belts were considered to be more suitable for younger unmarried girls and widows.

THE EASTERN REGION

JERICHO

The Jericho *thob* was mainly worn in Jericho and the eastern hills of Jerusalem and Bethlehem. The *thob* is extremely long, measuring up to three times a woman's height. The extra length was folded under the outer part of the garment and the extra layers secured around the waist with a belt.

The rationale for this length was both practical and "showy." The extra fabric protected the wearer from heat during the day and from cold at night. At the same time, the extra length indicated the wealth of the wearer, especially if the outer fold reached the hem of the dress.

The sleeves of the Jericho *thob* are very long and pointed, a feature that facilitated carrying objects. When a woman needed to work, she would either tie the sleeves over her shoulders and behind her back or over her head with a headband (*'asbeh*) or a patterned scarf.

Long dress (*thob*) of Jericho. OIM A35636A

THE SOUTHERN REGION

HEBRON AREA-BEIT JEBRIN

The Hebron region was very famous for its elaborate, variable, and distinct dresses. The *jallayeh*, which formed part of a bride's wedding attire in Beit Jebrin, is heavily embroidered in red with cross-stitch on the chest panel, sides, and back panel. The front from the waist to the hem is heavily decorated with colorful taffeta and other fabrics. The sleeves are also embroidered and adorned with atlas silk and taffeta fabrics in elaborate designs mainly in red.

Another dress of the region is called "heaven and hell" (*jinneh u nar*) from the name of a fabric made with selvedges of red and green silk. The chest piece is attached with cross-stitch and embroidered with various motifs, mainly the "ears of corn" (*sabal*) stitch. The sleeves of the dress are long and pointed and lightly embroidered.

The belt of the region is made of orange, yellow, or red atlas silk and is very similar to the belts worn by women in Bethlehem, Jerusalem, and Ramallah.

The head veil (*shambar*) of the Hebron region is made of black silk or crepe, heavily decorated on one end with a band of red embroidery. It has long silk tassels intertwined with gold thread. It was worn with the red band on the side so that when a woman covered her face the red embroidery would show.

Another head veil used in Hebron was called the *ghudfeh*. This head veil was made of cream-colored linen and was embroidered with beautiful multi-color thread with designs similar to those found on the dress.

Wedding dress (*jallayeh*) of the Hebron area. PHC 170

The headdress (*'iraqiyeh*) is a richly embroidered circular cap with coins its edges. Most *'iraqiyeh*s in Hebron had what is referred to as *laffayef*, long strips of heavily embroidered cotton or silk attached to each side of the *'iraqiyeh*, ending with a long wool tassel. A woman could wrap the *laffayef* around her braided hair to cover it, and the wool tassels made it easy to pin the wrapped braids to the main part of the headdress.

BIR SABE'

Bir Sabe' dresses were worn by Bedouin women from the southern part of Palestine. These women embroidered their dresses with the same brightly colored cross-stitch that was found throughout Palestinian villages, but they also created their own motifs highlighted by various geometric designs. They took the star, cypress tree, and palm motifs and used them in stylized forms. The mostly sateen fabric was always black or dark blue and was embroidered with very bright colors. The sleeves of this *Bedu* dress were triangular and pointed.

The chest panel of the Bir Sabe' *thob* was embroidered separately with cross-stitch before it was sewn onto the dress. In contrast to most dresses, which employ embroidered panels, the stitched decoration on the side, back, and even front of the Bedouin dresses are embroidered on the fabric itself.

Women of the semi-desert area used the lower back panel of the dress for self-expression and as a way of non-verbally communicating their desires and aspirations. A woman's identity was expressed mainly by the colors and designs used: blue threads for the young and for widows and red for married women. A widow would decorate

Dark blue sateen *thob*. Bir Sabe'. PHC 150

Bedouin face veil (*burqa'*) embellished with coins. PHC 151. Photo courtesy of Maha Saca

the back panel of her dark blue cotton cross-stitch embroidery with a few colored motifs indicating that she was seeking to remarry. Some women in this region embroidered motifs of little children on their dresses if they were interested in having children of their own. The importance of the decoration of the back panel in many regions is due to a conservative moral code. It was not considered proper to look at a woman's face or to stare at her from the front. However, watching her from behind as she passed by was considered more acceptable.

The head veil (*qun'ah*) was made of the same blue or black cotton material as the dress. The middle section of the head veil is lavishly embroidered with threads of red, orange, and green and with patterns similar to those on the dress. The embroidery on the veil ran vertically down the back of the head, ending around the level of the knees. The orientation of the embroidery on the head veil was executed in a way that visually connects it with the embroidered panel at lower back. Together, these elements create an unbroken colorful pattern that totally covers the woman from the back.

Underneath the head veil or shawl, Bedouin women wore a headdress or head cover (*sahliyeh*) made usually of atlas silk embroidered in cotton thread cross-stitch, covered with coins, beads, cowrie shells, and other items depending on the women's choice and what was available to her. Beautiful red wool tassels decorated the end of the *sahliyeh.*

The face decoration or veil (*burqa'*) was worn by married women and by girls after they reached puberty. The *burqa'* hung from a narrow band at the forehead and covered the nose, mouth, and neck areas, sometimes extending to the chest. A narrow band was usually embroidered and covered with coins, while the rest of it was covered with coins, beads, and semi-precious stones. For married women it was a symbol of wealth and status. The *burqa'* protected a woman's face from sand storms and sun and maintained her modesty.

The Bedouin of the southern part of Palestine wore belts (*hizam*) of wool that were locally woven and usually dyed red or pink. Some of these belts were simple with small colorful tassels, while others, made of goat hair, were more elaborate with highly ornamented tassels covered with beads and shells.

MEN'S CLOTHING

The traditional Palestinian ensemble for a man consisted of an undergarment (*qamis*), a coat (*qumbaz*), another type of shoulder mantel/cloak (*'abaya*), baggy pants (*shirwal*), and a belt (*hizam*). Some other items were later introduced, such as the wool overcoat and a short jacket. Although men's clothing was not as elaborate as women's clothing, it still gave clues to the age, status, and identity of the wearer.

A long undershirt or tunic (*qamis*) was worn throughout Palestine by both village and Bedouin men. This garment was made of cotton or very fine wool, usually an undyed white or cream. The length varies between knee and ankle depending on the region. The length of the sleeve also varied; some sleeves were long and pointed, while others were long and narrow.

The long coat (*qumbaz*) was worn over the tunic. The *qumbaz* had narrow sleeves, a round neckline, and was usually ankle length. It was often made of cotton, but the more expensive examples were made of Syrian silk. *Qumbaz* fabric was usually striped in very colorful combinations; some were multi-colored while others were much simpler utilizing only one or two colors. The white and blue striped *qumbaz* was the preferred color combination in many Palestinian villages. It is believed that in the late nineteenth–early twentieth century the color of the *qumbaz* reflected certain village or group associations, and men proudly wore their village colors in public ceremonies or festivals.

Wool cloak (*'abaya*) worn over a striped coat (*qumbaz*). OIM
A35660, A35648B

Man's overcoat (*'abaya*) of striped wool with metallic couching on the neck. OIM A35660

Man's tunic (*qamis*). OIM A35644A

Bedouin men would wear a *'abaya* (overcoat or shoulder mantle) made of hand-woven wool that was usually brown and white. Other more modern *'abayas* are made of a lighter material like fine wool or silk. These *'abayas* were often worn by wealthier men, and to add to the beauty of their overcoat, most had their *'abayas* embellished with a border of zigzag gold threads. During the cold winter months, instead of the *'abaya*, village and Bedouin men alike preferred to wear a knee-high sheep skin coat known as the *farwa*. The *farwa* was worn with the woolly part of the skin toward the body, which kept the wearer very warm.

The *shirwal* (baggy pants) was worn in the early and mid-twentieth century mainly in Palestinian villages. The *shirwal* was made of black, white, or dark blue cotton and it hung down in folds between the legs and fitted tightly around the ankle and lower leg.

Man's head wear composed of a wool cap (*tarbush*) wrapped in a silk fabric (*laffeh*). OIM A35645E–F

The two main headdresses worn by Palestinian men were the *tarbush* and *laffeh* and the *hata wi 'agal*. The *tarbush* is the Ottoman-style flat-topped conical hat made of red or burgundy felt with a black tassel usually made of silk. It was worn by men in towns during the period of Ottoman rule in Palestine. The Turkish *tarbush* was high and stiff while the Palestinian *tarbush* was made of softer material and was shorter and flatter. Men would wrap a piece of cloth (*laffeh*) made of Syrian silk or cotton around the *tarbush*. The color of the *laffeh* indicated the age, social standing, and religious affiliation of the wearer. For example, a *tarbush* wrapped with *ghabani* fabric was worn by village elders, while a *tarbush* with orange *laffeh* was worn by both Christians and Muslims in various villages throughout Palestine. The *tarbush* and *laffeh* were worn only by adult men.

The *hata wi 'agal* was originally the headdress of Bedouin men, but it became more widespread in the early twentieth century as the turban went out of style. The *hata* is a large square piece of white cotton or silk. Men would fold the *hata* into a triangle and secure it to their head using an *'agal*, a circle of rope made from goat hair or wool. The most common color for the *'agal* is black, but some elaborate *'agals* were made of brown or black camel hair and incorporated gold thread into the weave.

Most men wore a small cotton skullcap (*taqiyyah*) underneath the *hata*. The *taqiyyah* helped to keep the head dry on hot days and warm on cold winter days.

Men's footwear was simple and made locally. Most village and Bedouin men during the nineteenth century preferred to walk barefoot, believing that wearing shoes all the time was an unclean practice. When shoes were worn, they were made of brown or red leather; some Bedouin men wore sandals.

Belts varied in thickness and were made of a variety of materials, including wool, cotton, silk, or leather. Most village and Bedouin men preferred to wear cloth belts, called *shamleh*, wrapped around their waist. For those who could afford it, the *shamleh* was made of atlas silk, but in general the belts were made of soft wool and cotton. Bedouin men also wore belts decorated with shells and embroidery and had multiple hooks that allowed the wearer to attach tools and personal items such as a tobacco pouch or a small dagger. Later, leather belts (*hizam*) became more popular and were used by both younger village and Bedouin men. But most older men preferred the *shamleh*.

Man wearing striped coat (*qumbaz*), cloak (*'abaya*), and cotton skull cap (*taqiyyah*). Circa 1924–1946. Library of Congress, Prints and Photographs Division, LC-DIG-matpc-13714

PALESTINIAN JEWELRY

Jewelry was an essential part of a Palestinian woman's possessions. From the nineteenth century to the mid-twentieth century, women adorned themselves with silver bracelets, necklaces, rings, nose rings, and chokers. Most of the jewelry was made locally by professional silversmiths, many of whom were Christians and Jews who passed the craft down though their family. The jewelry was also heavily influenced by silversmiths from Egypt, Syria, and other areas.

Some Palestinian jewelry was regionally distinct, but many villages within the same immediate vicinity shared styles and jewelers. Jewelry was not simply an adornment, but also possessed certain associations and social meanings. For example, Bedouin and some village women from southern Palestine wore fat tube-like silver jewelry sometimes called the "cucumber" in which they inserted verses copied from the Quran. These pieces of jewelry were worn for protection and good luck. Other necklaces composed of oval or triangular pieces of silver were engraved with the name of God for protection. Amulets (*hijab*) were also very popular because they were thought to protect the wearer from evil, envy, or sickness.

The *iznaq* was worn by married women in the Bethlehem area. This piece is made of a large silver chain that hung from the *shatweh*, the married woman's headdress. The *iznaq* could be double-chained and from it hung an additional seven chains (*sabe' erwah*) or five chains (*khams erwah*). These chains were usually adorned with silver roses and at the lower end where all the chains met hung a Maria Theresa coin. Some Christian women substituted a cross for the Maria Theresa coin.

Silver choker (*bughmeh*) with chains ornamented with early nineteenth-century Ottoman coins. OIM A35638G

Silver chin-chain (*iznaq zarad*) that secured the *saffeh* headdress. The large Maria Theresa coin, originally minted in Austria, was widely accepted in parts of the Middle East until the mid-twentieth century. PHC 136

Group of silver bracelets. Left: *haydery*, middle: *masriyeh*, right: *dumlug*. PHC 149, 130, 155

The popular *haydery* bracelets were produced locally in Bethlehem, and they were worn by women from that town and the surrounding villages as far south as Hebron. These bracelets were produced by four or five different silversmiths who lived throughout Palestine. Some of these silversmiths are known to us today because they stamped their names on the front of the bracelets so their work could be instantly recognized. Most women wore these bracelets in pairs, one on each wrist, but some preferred wearing two per wrist. Some women mixed styles to demonstrate their wealth. In the 1920s, new designs of jewelry made of gold flooded the markets and women were drawn to the new styles, eventually abandoning the locally made silver jewelry.

Other beautifully worked silver jewelry worn by women throughout Palestine in the nineteenth and twentieth centuries include silver chokers (*kirdan* and *bughmeh*), which were very popular, especially in southern Palestine; ankle bracelets (*khilkhal*); silver hair pieces (*karamil*); and silver pendants, especially amulets and nose rings (*shnaf*). In 1948, after the establishment of the state of Israel, the production of this silver jewelry stopped and many Palestinian women sold their possessions in order to survive. Most of the silver jewelry had disappeared by 1967 when the West Bank and Gaza were occupied.

CLOTHING IN PALESTINE POST-1948

The creation of the state of Israel in 1948 disrupted life in Palestine and the surrounding region. Many villages were destroyed or occupied, and many Palestinians became refugees in the West Bank and Gaza and in neighboring countries, especially Jordan, Syria, and Lebanon. In the following decades, major transformations took place in clothing styles and more specifically in the art of embroidery.

There are many reasons for the disappearance of the traditional dress and the art of embroidery as a whole. With the forced dislocation, refugees focused on surviving in the camps and finding ways to provide for their families. Most of the Palestinians became poor and could not afford to buy the traditional materials needed to embroider elaborate dresses. The main weaving centers, like the one in Mejdel, became part of the newly established state of Israel and Palestinians no longer had access to these centers or their products. It also became easier and cheaper to purchase imported fabrics.

The regional distinctions in embroidery and dress styles lost their social significance; women in the refugee camps focused on a shared or common identity as "Palestinians" rather than focusing on regional distinctions. In the 1970s, for example, women embroidered their dresses with motifs that they thought were fashionable or beautiful, ignoring regional distinctions in motif, pattern, and fabric. The many new dress designs had the same

Modern dress (*thob*) decorated with Palestinian flags, images of the Dome of the Rock, and the word "Palestine." PHC 173

basic structure and characteristics of pre-1948 dresses, but they were much simpler in style and decoration, and hence were less costly and time-consuming to produce.

The two main new types of dresses produced are known as the four branch (*arba' agruq*) or six branch (*sit agruq*) dress, based on the number of branches (*'arq*); and the *maris* dress, which some refer to as the *shawal* dress. The six or four branch dresses refer to the number of vertical embroidered stripes on the front and back of the skirt starting at the waist and going down to the hem. These embroidered stripes had various designs ranging from elaborate foliage and floral motifs to complex human and animal figures and motifs. The chest panel on the six branch dress was no longer embroidered in regionally distinct motifs, and it was enlarged and embroidered with a mix of motifs and border designs. The lower back panel was still considered an important part in these modern *thobs*, especially the *maris thob*, but the area that is covered with embroidery is much smaller and usually matches the design on the chest panel. The four and six branch *thob* are still worn today throughout the region.

The *maris thob* is made of heavy linen. It was mainly mass produced in the refugee camps and was sold in an uncut form so the purchasers could assemble it. Women still embroider on the fabric itself using a trace with "traditional" motifs as well as new patterns that show European influence. The style of the dress was also Western, especially the narrow, slim cut.

In the 1960s, black sateen or colorful acrylic fabrics replaced the more traditional handwoven linen and cotton fabrics, especially in the dresses made in refugee camps. Decorative motifs became "European" and the thread was made of cotton rather than silk. Most importantly, in the 1980s, some handmade embroidery gave way to machine-made decoration. All these changes reflect the harsh economic, social, and political situation that the Palestinians were facing in the West Bank and Gaza and the Diaspora especially of the 1960s and the 1970s.

TRADITIONAL EMBROIDERY TODAY

In the decades since the 1950s, embroidered dresses and the art of embroidery as a whole began disappearing from the living memories of Palestinians, especially among the younger generations. The older women still wore the dresses and embroidered them, but the younger women were not interested in this art and viewed embroidered dresses and even embroidered pieces in the home as an outdated fashion fit only for the elderly.

There were attempts to revive and preserve this tradition. Older Palestinian women from various refugee camps continued embroidering their dresses as well as embroidering pieces for the home. In the 1980s, international funding agencies provided women with small grants to mass produce and sell their embroidery. This was an effort to improve the economic situation of families living in the refugee camps and develop embroidery as an art separate from its traditional context and meaning. Many new colors, designs, motifs, and patterns emerged to suit the target markets in the West. Decorative ornaments with Christmas themes, purses, and book covers in colors that appealed to the Western market started to appear. These efforts directly and indirectly helped to maintain the art of embroidery and the preservation of some kind of embroidered clothing, even if these were not strictly traditional. Such efforts were viewed by many as a way to revive Palestinian cultural heritage in the West Bank and Gaza and more importantly in the Diaspora. In the late 1980s and early 1990s, Palestinian women began to embroider symbols of their national identity onto their clothing as well as on various articles such as shawls and wall hangings. Images of the Dome of the Rock in Jerusalem, the colors of the Palestinian flag, and the word "Palestine" embroidered on garments and other items became popular symbols of their homeland and national identity. This method of personal and national expression was enhanced during the first Intifada.

Recent years have seen a renewed interest in the preservation and promotion of traditional Palestinian clothing and embroidery, encouraged and sponsored by groups such as the Palestine Heritage Center, Bethlehem, Palestine, led by Maha Saca (kneeling, center). Photo courtesy of Maha Saca

In an effort to revive the interest of the younger generation in traditional embroidery, heritage institutions and centers, like the Palestinian Heritage Center, commissioned traditional patterns and motifs to be embroidered on modern clothes. The practicality and beauty of these clothes attracted the interest of young Palestinian women. Today, many wear embroidered shirts, skirts, or pants on a daily basis as well as at special events; they wear them for their beauty and meaning.

From the late 1990s to the present, this movement of national- and self-expression through embroidery on dresses and other articles continues as another younger generation expresses a renewed interest in the value and meaning of the embroidery. This revivalist movement can be seen today not only in the West Bank and Gaza, but also in the United States, Jordan, Europe, and even Australia. Many women are returning to the motifs, colors, and fabrics of the nineteenth and early twentieth centuries. They are discouraging the use of machine-made, mass-produced embroidery and they are placing a high value, both emotional and monetary, on handmade embroidered items and dresses that truly reflect traditional Palestinian designs and colors.

Today, in many cities such as Ramallah, Bethlehem, and Jerusalem, well-to-do women are wearing handmade embroidered dresses with traditional and non-traditional motifs and patterns. They consider these dresses not only to be fashionable, but also to represent an identity that Palestinian women are trying to preserve, an identity that connects them to their ancestors and their land. Among the centers in the West Bank and Gaza and the Diaspora that are trying to preserve this heritage are the Palestinian Heritage Center in Bethlehem, the In'ash al Usra in Ramallah, Dar el Tifl al Arabi in Jerusalem, Kawar Arab heritage collection in Jordan, the Palestine Costume Archive in Australia, and the Palestinian Heritage Foundation in the United States, to name just a few.

Today a wonderful tradition is starting to emerge in cities in the West Bank and Gaza. Imagine you are going to a wedding in Bethlehem — whether Christian or Muslim — and you will observe the attempt to revive the traditional Palestinian dress and the art of embroidery. As in the West, a bride will wear a beautiful wedding dress, enjoy the wedding reception, and dance all night with family, friends, and guests. But many young Palestinian brides are starting to wear the traditional wedding dress — mainly the *thob al malak*, the bride's dress of Bethlehem — the night before the wedding, the henna night, or even the day of wedding between the various ceremonies. Many young brides today choose to have a traditional dress embroidered for her special day, or if she does not have time or cannot afford it, she might rent it for the occasion.

Throughout the years, Palestinian embroidered dresses and their accessories have undergone many changes. Yet, Palestinian women today are determined to keep this tradition alive, a tradition that allows them to comfortably and proudly express their identity as women and as Palestinians. It is hoped that these garments demonstrate the beauty, technical achievement, and tremendous diversity of regional clothing in mid-nineteenth and early twentieth century Palestine and remind people of the role that clothing and embroidery play in developing and maintaining identity in Palestinian culture, and in cultures all around the world.

EXHIBIT CHECK LIST

All objects, unless otherwise indicated, date to the late nineteenth and early twentieth centuries.

Abbreviations: OIM: Oriental Institute Museum; L followed by digits: loan to Oriental Institute Museum; PHC: Palestine Heritage Center, Bethlehem

---

COMPONENTS OF A PALESTINIAN DRESS

Chest panel (*qabbah*): PHC 304; Sleeve (*kum*) with decorated cuff (*sfifeh*): PHC 302; Skirt side panel (*benayiq*): PHC 307

COASTAL REGION

Jaffa-Beit Dajan
Dress (*jallayeh*): OIM A35643A; Headdress (*smadeh*): L-436.01

Gaza
Dress (*thob*): PHC 165; Beaded hat (*takiyat shabak*): PHC 166; Beaded shawl (*guddeh*): PHC 167

CENTRAL REGION

Ramallah
Group 1
Dress (*thob*): OIM A35635A; Shawl (*khirkah*): OIM A35635B; Headdress (*saffeh*): OIM A35633E; Chin-chain (*iznaq zarad*): PHC 136; Belt (*zunnar*): PHC 201; Choker: OIM A35640D

Group 2
Dress (*thob*): OIM A35633A; Shawl (*khirkah* or *shal*): PHC 135; Chin-chain (*iznaq zarad*): PHC 132; Belt (*zunnar*): PHC 202

Bethlehem
Dress (*thob al malak*): OIM A35640A; White veil (*khirkah*): PHC 171; Jacket (*taqsireh*): OIM A35652; Belt (*shamleh*): PHC 113; Headdress (*shatweh*): OIM A35640E; Necklace (*iznaq kahms 'erwah*): PHC 112; Hat (*shatweh*) OIM A35640E, PHC 112; Necklace (*bughmeh*): OIM A35638G; Sleeveless jacket (*bisht*): OIM A35646B; Headdress: L-436.01; Headdress: L-436.03; Child's hat: PHC 306; Tassels: PHC 310

Jerusalem
Dress (*thob*): PHC 119; Circular cap (*qub'a*): OIM A35643C; Veil (*khirkah*): OIM A35645G; Belt (*hizam*): OIM A35643B

EASTERN REGION

Jericho
Dress (*thob*): OIM A35636A; Crescent moon silver necklace (*hlalat*): PHC 138

SOUTHERN REGION

Bir Sabe'
Dress (*thob*): PHC 150; Face veil (*burqa'*): PHC 151; Shawl (*kun'a*): PHC 153; Belt (*hizam*): PHC 203

Hebron Area-Beit Jebrin
Dress (*jallayeh*), PHC 170; Shawl (*shambar*), PHC 141; Headdress ('*iraqiyeh* with *laffayet*) L-436.04; Headdress ('*iraqiyeh*): L-436.03; Bracelets: *masriyeh*: PHC 130, *haydery*: PHC 149; Silver bracelet: PHC 155

MALE COSTUMES

Group 1
Undergarment (*qamis*): OIM A35639A; Coat (*qumbaz*): OIM A35648B; Cloak ('*abaya*): OIM A35634A; Belt (*shamleh*): OIM A35682B; Skull cap (*taqiyyah*): OIM A35637D; Head cover (*hata*): PHC 122; Head cover rope ('*agal*): OIM A35639E

Group 2
Undergarment (*qamis*): OIM A35644A; Coat (*qumbaz*): OIM A35641A; Cloak ('*abaya*): OIM A35660; Belt (*hizam*): OIM A35644E; Headdress (*tarbush* and *laffeh*, *keffiyeh*): OIM A35645E-F

MODERN SYMBOLS OF IDENTITY

Dress (*thob*): PHC 173

GLOSSARY

'abaya: loose-fitting sleeveless overcoat that was draped over the shoulders, worn by men and women.

Abu Mitain: cotton dress fabric in indigo blue linen with wide violet-red silk stripes and smaller red and green silk stripes on each selvedge of the dress, woven in Mejdel for Gaza and Ashdod dresses.

Abu Rukbeh: Gaza dress made of indigo blue linen with heavy silk embroidery.

'agal: circle of rope made from goat hair or wool that secures a man's fabric head cover.

'ambar: everyday Bethlehem *thob.*

'arq: branch.

asawri: Syrian silk fabric horizontally striped in red and yellow or in yellow and black.

'asbeh: headband.

atlas: silk fabric striped in red and yellow.

benayiq: triangular skirt side panels.

biltajeh: indigo blue linen fabric with colored borders used for *thobs* in the coastal region, woven in Mejdel.

bisht: sleeveless wool red coat/mantle of Bethlehem. The term could also be used for men's wool coat.

bughmeh: silver choker with chains, part of the bridal jewelry.

burqa': face decoration (veil) worn by married women.

couching: technique of decoration by overlaying gold or silver cord on fabric in various floral designs and securing it with small stitches. Mainly done by embroiderers of Bethlehem.

diyal: hem/border of a *thob.*

farwa: knee-high sheep skin coat.

ghabani: white silk or natural cotton fabric embroidered with gold or white chain stitch embroidery from Aleppo, Syria.

ghudfeh: head veil.

hata: man's large square head cover made of white cotton or silk.

hata wi 'agal: headdress for men.

haydery: silver bracelet.

heremzi: colored silk material made mainly in Syria.

hijab: an amulet in the context of garments; an embroidery motif in a form of a triangle.

hijjer: front part of the skirt of a *thob.*

hizam: belt.

hlalat: crescent moon silver necklace.

'iraqiyeh: richly embroidered headdress adorned with coins, worn mainly in the Hebron area.

irdan: long pointed triangular sleeves of a *thob.*

iznaq: silver ornament chain that hangs from a headdress.

iznaq zarad: silver ornament chain that hangs from the Ramallah headdress.

jallayeh: heavily embroidered women's long wedding dress or coat open in the front, worn in the Galilee area, Ramallah, and Hebron.

jinneh u nar: "heaven and hell" cotton fabric — also the name of a dress — with selvages of red and green silk. The red represents hell and green represents heaven.

joseh: extension on chain of Ramallah headdress.

kankhel 'ali: tall palm.

karamil: hair ornaments.

keffiyeh: head scarf worn mainly by men.

khilkhal: ankle bracelets.

khirkah: shawl and generic name for a woman's veil.

kirdan: silver choker.

kum: narrow long or short sleeves of a *thob*.

kun'a: *Bedouin* veil; see *qun'ah*.

laffayef: long strips of heavily embroidered cotton or silk attached to each side of an *'iraqiyeh* headdress, ending with a long wool tassel. Women rolled them around their hair.

laffeh: length of fabric wrapped around a *tarbush*.

malak: "royal" fabric of linen with a higher percentage of silk than other fabrics.

manajel stitch: one of the two main stitches employed in traditional Palestinian dressmaking. The other is the *sanabel* stitch.

maqta': length of dress fabric of Gaza 35 centimeters by 7 meters.

maris: see *mawaris*.

masiriyeh: bracelets.

mawaris (*maris* singular): two vertical bands of embroidery on the front side of a dress that frame the center of the skirt.

najmah kin'aniyah: eight-point or Bethlehem star motif.

nol: traditional loom.

nujum: star patterned embroidery motif.

qabbah: chest panel of dress, usually heavily embroidered.

qamis: man's undershirt or tunic.

qasab: use of gold or silver cord in embroidery.

qos: arch-shaped embroidery motif famous on Ramallah's chest panel.

qub'a: a small circular embroidered headpiece worn in Jerusalem and Bethlehem.

qumbaz: men's long coat with one side crossing over the other. The *qumbaz* is usually made of striped material and is worn belted.

qun'ah: head veil worn by the Bedouins. The long large veil is made of blue or black cotton fabric with the middle part embroidered with reds, orange, and green threads.

radah: shoulder piece or yoke of a *thob*.

reesh: embroidery pattern, also called "feathers." This pattern is used for borders.

rumi: linen fabric.

sa'at: "watch"-shaped embroidery design.

sabal: herringbone "ears of corn" embroidery stitch.

saffeh: woman's horseshoe-shaped Ramallah headdress with coins on the front part of the headdress forming a halo around the face.

sahlyieh: headdress or head cover made usually of atlas silk embroidered in cotton thread cross-stitch, covered with coins, beads, cowrie shells, and other items worn mainly by Bedouin women.

sanabel stitch: one of the two main stitches employed in traditional Palestinian dressmaking. The other is the *manajel* stitch.

saru: cypress tree.

sfifeh: cuff of a dress' sleeve.

shal: large silk shawl.

shambar: thick embroidered head veil, usually black with thick red embroidery on one side.

shamleh: belt worn by Bethlehem women. Made of locally woven wool fabric in pink or blue.

shatweh: elaborately embroidered Bethlehem area headdress. The front part is adorned with bride's wealth of coins and coral.

shawal: a name given to modern *thob*s made after 1948.

shinyar: lower back panel of a *thob.*

shirwal: baggy pants that narrow at the ankle, usually embroidered on the lower part near the ankle.

shnaf: nose ring.

smadeh: distinctive horseshoe-shaped Ramallah and Jaffa headdress with a row of coins attached to its front. The term is also used for the northern coastal area headdress.

ta'er al finiqe: bird of life.

tahriri: couching in silk.

takiyat shabak: beaded cap of the Gaza region, made of colorful, especially blue, beads.

taqiyyah: man's small cotton skullcap worn to absorb sweat.

taqsireh: short jacket of Bethlehem's bridal outfit, heavily embroidered with silk and gold and silver threads.

tarbush: high stiff burgundy cap for men, introduced from the Ottoman empire, usually made of felt.

thob: generic term for all women's dresses.

thob al malak: "royal" bridal dress of Bethlehem. So popular it was worn by women from other surrounding villages and towns.

thob al malak al ikdari: "green royal dress" of Bethlehem, made of red, green, and indigo striped fabric with a high percentage of silk.

toq: horseshoe-shaped band of Maria Theresa coins placed on top of a *saffeh.*

zrad: chin-chain.

zuniyet: thob for Gaza made of indigo blue linen striped in magenta and light green.

zunnar: a belt. This term is used for both men and women's belts. A *zunnar* could be made of leather or a large folded rectangular cloth.

BIBLIOGRAPHY

Abu 'Umar and 'Abd al-Sami. *Traditional Palestinian Embroidery and Jewelry*. Jerusalem: Al Shark Arab Press, 1986.

Allenby, Jeni. *Portraits without Names: Palestinian Costume*. Canberra: Palestine Costume Archive, 1996.

Amir, Ziva. *Arabesque: Decorative Needlework from the Holy Land*. New York: Van Nostrand Reinhold, 1977.

El-Khalidi, Leila. *The Art of Palestinian Embroidery*. London: Saqi Books, 1999.

Grutz, Jane. "Woven Legacy, Woven Language." *Aramco World* 42 (1991): 34–43.

Kana'neh, Sharif, et al. *Palestinian Folk Costume*. El Bireh: Family Relief Society, 1982 (in Arabic).

Kawar, Widad. "Traditional Costumes of Jordan and Palestine." *Arts and the Islamic World* 2 (1984): 41–42.

Meir, Cecilia. *Crown of Coins: Traditional Headdresses of Arab and Bedouin Women*. Tel Aviv: Eretz Israel Museum, 2002.

Munayyer, Hanan Karaman. "New Images, Old Patterns: A Historical Glimpse." *ARAMCO World* 48 (1997): 2–11.

Muzayyin, 'Abd al-Rahmān. *Mawsū'at al-Turāth al-Filasṭīnī*. Manshūrāt Filasṭīn al-Muḥtallah, 1981 (in Arabic).

Palestinian National Costume - Preserving the Legacy. Palestinian Heritage Foundation, 1995. Video.

Rajab, Jehan. *Palestinian Costume*. New York: Kegan Paul International, 1989.

Society of In'ash El-Usra. *Palestinian Embroidery*. El Bireh, 1996.

The Revival of Palestinian Traditional Heritage: Documentation of "Inash" Embroidery over Thirty Years Based on Traditional Old Palestinian Motifs Adapted to Modern Designs. Beirut: Inaash, 2001.

Weir, Shelagh. *Palestinian Costume*. London: British Museum, 1989.

———. *Palestinian Embroidery: A Village Arab Craft*. London: British Museum, 1970.

Weir, Shelagh, and S. Shahid. *Palestinian Embroidery: Cross-stitch Patterns from the Traditional Costumes of the Village Women of Palestine*. London: British Museum, 1988.

Interview:

Saca, Maha, Palestinian Heritage Center, July 2006

Websites:

http://www.palestinecostumearchive.org

http://www.palestineheritage.org

http://www.palestinianheritagecenter.com

http://www.inash.org/index.html